AFRICA'S
# BIG CATS
AND OTHER CARNIVORES

## Nigel Dennis

SUNBIRD
PUBLISHING

First published 2001
2 4 6 8 10 9 7 5 3 1
Sunbird Publishing (Pty) Ltd
34 Sunset Avenue, Llandudno, Cape Town, South Africa
Registration number: 4850177827

**Publisher** Dick Wilkins
**Editor** Brenda Brickman
**Designer** Mandy McKay
**Production Manager** Andrew de Kock

Reproduction by Unifoto (Pty) Ltd, Cape Town
Printed and bound by Tien Wah Press (Pte) Ltd, Singapore

**ISBN 0 624 0397 06**

TITLE PAGE *The adult male lion is the primary defender of the pride's territory.*

LEFT *Lion territories are protected by a single male, or, more usually, a coalition of adult males, creating a safe environment in which the lionesses can raise cubs.*

BELOW *Spotted hyena cubs are generally restricted to the den and its immediate surrounds until they are about a year old.*

OPPOSITE *A cheetah, swiftest of all cats, surveys the open veld for prey.*

# Africa's Big Cats and other Carnivores

In many African environments big cats and other carnivores are often found living alongside one another, resulting in a complex set of interactions and effects that are not always easy to tease apart and understand. Undoubtedly, however, lions are the lords of the bush. Aside from their sheer, brutish power, the strength achieved through family alliances leaves them unmatched as a group. Nonetheless, spotted hyenas will, albeit rarely, challenge lions for prey in a display that is at the same time riveting and shocking to witness. The roots of this intense rivalry may lie in the fact that hyenas and lions favour similar prey – it is certainly not because spotted hyenas are unable to hunt successfully themselves.

Large carnivores are not restricted in their choice of prey. Even small elephants and adult rhinos or hippos that are past their prime are no match for a large pride of determined lions. Buffalo and giraffe, occurring in large herds, are eagerly sought after prey species, although adults are easy pickings only to large prides or male coalitions, whose combined strength is necessary to subdue them. Conversely, hungry lions are not averse to thrashing at a small acacia bush until it expels its hundreds of red-billed queleas, and ludicrously slap at the diminutive birds as they take flight. Although lions will catch such prey from time to time, they will, more commonly, select the medium-sized game found in their home ranges. In many areas this requirement is met in the form of wildebeest or zebra, but other buck such as gemsbok, waterbuck or kudu form the bulk of a lion's diet in habitats where these latter-mentioned species are most prominent.

In regions where only smaller antelope and warthog exist, lions will hunt more frequently in order to match their necessary requirements for meat.

Lions can appear strangely brutal, especially with regard to killing fellow carnivores, where they will expend considerable energy in chasing down and killing such relatively tiny competitors as jackals or mongooses. They will, however, also kill any of the larger predators should the opportunity arise. The reasons for this are unclear, but seem to be based on reducing the competition.

Leopards, on the other hand, do regard other carnivores as prey, and it is not unusual to see these dappled cats with a treed and partially eaten jackal or civet kill. Cheetahs too are known to catch and eat smaller carnivores such as jackals and bat-eared foxes, albeit on relatively rare occasions.

Ecological separation among these large carnivores in terms of selection of prey species, the age or class of

*ABOVE Wild dogs are diurnal and cooperative hunters, and rarely, if ever, rely on scavenging for their existence.*

creature generally hunted and the time of day in which they hunt, are some of the main mechanisms by which large carnivores attempt to reduce competition with each other. Leopard, cheetah and wild dog all tend to hunt similar prey, although the leopard is far more adaptable and kills a much wider variety of animals. Leopards focus their hunting efforts on smaller antelope species such as springbok, impala, and gazelle, as well as reedbuck, bushbuck, duiker and steenbok. Where the cheetah is almost exclusively a mid-morning or afternoon hunter, the leopard hunts mainly at night, or in the magical hours of transition from day into night. Furthermore, leopards tend to hunt under denser cover, from which they can stalk up close to their prey, whereas cheetahs favour hunting in open environments where, with lightning speed, they can chase down their quarry over a couple of hundred metres.

*BELOW Leopards lead predominantly solitary lives, although there may be contact between neighbours.*

The social structure of these big cats and other large carnivores is as interesting and complex as their feeding habits. From the pride of related lionesses that support one another in the perilous period after their cubs have been born, to the typically cat-like solitary existence of the leopard, lies a fascinating range of complex social behaviour. It is now widely accepted that lions – unique among cats in their close family association, even into adulthood – live in prides not necessarily because of the increased chance of successful hunting in larger groups, but more so for the collective defence and nurturing of their cubs.

It is a popular misconception that a pride of lions is defended by a single pride male. In reality, prides are usually defended by coalitions of mainly two, but up to six, adult males. Most lionesses in the pride give birth at around the same time, resulting in a cohort of related cubs that needs to be protected against foreign and potentially infanticidal males who may kill the cubs. Inevitably, rogue males – looking to maximise their

BELOW *The social bond among lions is very strong, and even when resting they are constantly in close contact.*

reproductive output – will try to take over a pride in the absence of the adult or pride males, who fulfil the main defensive role. It is in these rare but critical times that the lionesses must band together and defend their cubs.

Even in a species seemingly as solitary as the leopard, the territorial male – whose range generally encompasses that of a few females – will often be in close proximity to a female in the period just after she has given birth to cubs, and he will drive off rival males. It may be purely accidental, but in defending his own territory the male leopard will keep other intrusive males that might harm his cubs at bay. A female in oestrus will encourage the male to mount her, and, as soon as he does, she will snarl and swipe a paw at him; this behaviour is temporary, and she will, after the show of aggression, allow him to mate with her.

Leopards are unique in that they will haul prey, up to twice their own weight, up a tree, and they are unusual in that they will eat carrion, usually returning nightly to their cache to feed.

Female cheetahs are as solitary as leopards, but several lone females often live in adjacent territories, which are also usually defended by a coalition of two to three adult males. A female cheetah that is about to come into oestrus is located by males from the smell of her urine. The male's initial advances are aggressively rebuffed, and he will respond with similar aggression. The male will persist, however, and continue to consort with the female until, after some two weeks of courtship, the female becomes receptive.

In all three of Africa's large cats, even the social lion, males will vie for mating rights with receptive females. Cheetahs fight aggressively until one of the males backs down. With lions, a short and extremely aggressive encounter between coalition males at the time of the lioness's receptiveness decides right of

ABOVE *Caracals are successful hunters, and will eat prey as small as dassies and as large as young antelope.*

access to her. Although the victorious male will consort closely with the lioness, in some instances she will escape his attentions and seek out the vanquished male with which to mate. Both leopards and lions mate frequently over a short period of time.

A fine line distinguishes Africa's 'big cats' from its 'smaller cats', and the largest African member of the genus *Felis* – the caracal – is more like a small 'big cat' than a big 'small cat'. An adult male caracal can reach a relatively hefty 17 kilograms (37 pounds); there is not much difference in weight between it and a small adult female leopard, which can weigh as little as 20 kilograms (44 pounds). Indeed, in many areas where the two species occur side by side they prey on similar species. The powerful hindquarters of the caracal propel the animal into spectacular leaps in an effort to pluck wildfowl on the wing right out of the air.

The caracal is often referred to as a lynx, probably due to physical similarities such as the long, tufted ears. In fact, there are several differences between the two, most notably the lynx's spotted and barred coat.

The serval, also of the *Felis* genus, is widely distributed throughout Africa's open grassland and savanna ecosystems. Long-legged and prettily marked, this small cat is elegant and agile. Like the caracal, it too is capable of making spectacular, high-curved leaps, and can even change direction in mid-air. The serval is a solitary animal that hunts – anything from snakes to insects – mostly during the cooler hours of the day, and generally inhabits long grass close to water. The tall grasses provide the serval with cover, protecting it from predation by other carnivores, in particular spotted hyenas and wild dogs.

*BELOW Spotted hyena cubs are suckled for up to a year, longer than any other carnivore.*

Spotted hyenas hunt mainly medium-sized game, although they tend to catch a greater number of juveniles of these species than lions do. When large prey species such as adult buffalos are in a weakened condition, a hyena pack will effectively hunt and kill them. Spotted hyenas are incredibly successful hunters, but will scavenge when the opportunity presents itself, particularly in areas where the ratio of lions to hyenas is greater. Usually, spotted hyenas significantly outnumber lions in areas where huge concentrations of migratory game such as wildebeest and zebra occur, such as the Ngorongoro crater or the Savuti marsh. In areas such as the Kruger National Park, lion numbers are high, as are the densities of ungulates, and although spotted hyenas do hunt occasionally, carrion is so plentiful that it provides for most of their dietary needs. In an arid area such as the Kalahari, where spotted hyenas occur at lower densities than lions, there are insufficient scavenging opportunities, and here they are almost exclusively hunters, mainly seeking out juvenile gemsbok as prey. Spotted hyenas will also, on occasion, chase down an adult gemsbok. The hyenas will run the gemsbok down, until, totally exhausted, it will inevitably back into a bush to protect its hindquarters. The stealthy hyenas will tear open its soft underbelly, skilfully avoiding the animal's rapier-like horns. Spotted hyenas will consume virtually ever part of a kill, including teeth, hair, bones and hooves!

They have a fascinating and unusual social system, whereby females are dominant over males, and single males are subordinant even to the cubs of high-ranking females, and the pack is usually led by an alpha female. It is the more heavily built females that band together to fight the territorial battles, and to hunt large prey. Coupled with the fact that a group of females retains

*ABOVE Juvenile cheetahs sport a fuzzy 'cape' of silver-coloured fur; only faint spots are initially visible on the coat.*

a territory for many generations, and the males are the transient sex, it is not surprising that spotted hyenas have a female-led society. In an apparent reversal of gender, the female clitoris has been hugely enlarged and mimics a penis in most ways, including being erectile. When encountering one another, clan members have an elaborate greeting ceremony in which a subordinate hyena initiates the greeting, and is more likely than a dominant animal to extend its 'penis'.

Although brown hyenas are mainly observed singly, like their spotted cousins, they also live in clans of up to 10 individuals that share a den in a burrow or cave. However, unlike spotted hyenas, they will carry pieces of meat and bone back to their cubs at the den, waiting for days if necessary for lions to vacate a kill so that they can scavenge the remains.

When two clan members meet, the younger member will present its anal region for sniffing, with head held low and ears flattened, while whining like a cub begging for food. Greetings between brown hyenas of the opposite sex and from different groups usually follow the same routine, and are peaceful. However, greetings between neighbours of the same sex are accompanied by ritualised threat displays, where the hyenas' manes stand erect in a false indication of size; the aggression may escalate to neck-biting, where blood is often drawn. Within a brown hyena clan there is a weak dominance hierarchy apparently based on age.

The brown hyena seems rather meek and mild in comparison with Africa's other bigger, faster and more aggressive carnivores. However, the brown hyena leaves these 'superstar' carnivores standing when it comes to endurance and perserverance. Brown hyenas are able to walk for incredibly long distances, and can smell out even the oldest bones or the smallest carrion remains. Like a shaggy ghost in the night, a brown hyena will disperse from the communal den in the early evening and begin its lonely wanderings into the night in search of a meal.

Wild dogs tend to hunt in the early morning or late afternoon, and are able to exhaust their prey in

*BELOW Although some nomadic lionesses do exist, a solitary lioness is not a common sight.*

extended chases, often over several kilometres. Although wild dogs seem to favour a wide range of habitats, when living in the same environment as lions they tend to select the areas of lower lion density in order to avoid predation. In the Kruger National Park, where lion densities are higher due to the wide availability of zebra, wildebeest and buffalo species, particularly on the grassy plains in the eastern reaches of the park, wild dogs have naturally selected the wooded, western areas in which to pack; this even though their favoured prey – impala – occurs in higher densities in the more eastern areas. Generally speaking, wild dogs do not kill other carnivores, but will chase and harass those that they encounter, and it is not an uncommon sight to see a leopard trapped in a tree and surrounded by a pack of baying wild dogs.

Aside from having to contend with predatory lions, wild dogs are continually harassed by spotted hyenas, which are content in the knowledge that if they follow their smaller competitors, they will invariably end up with an easy meal. While wild dogs will, to some extent, tolerate their presence, they regularly and aggressively chase after and even mob a single, persistent hyena, and large packs will chase off small groups of scavenging spotted hyenas. A solitary hyena will usually try to find cover from mobbing wild dogs by tucking in its hindquarters – the target of the dogs' vicious nips – and skulking off into a bush, or scaling a termite mound and sitting on its targeted haunches.

Wild dogs have their own unique, close-knit social structure, which is one of the highest forms of mammalian social organisation. As with most canids there is an alpha pair, which is generally the only pair in the pack to reproduce. The rest of the pack consists mostly of the offspring of the current or previous alpha pair. Wild dogs are unusual in that the females are

*ABOVE Endemic to southern Africa, the brown hyena is a shy and solitary species.*

more likely to disperse from the pack than males; most males remain in their natal pack and become helpers while they wait for the loss of the dominant male, which will perhaps allow them an opportunity to breed. Large litters of pups often result in the dispersion of big groups of young adults of the same sex; a new pack is formed when they meet a group of the opposite sex.

The big cats and other large carnivores of Africa are indeed a highly complex and fascinating group. At once ruthless and savage towards competitors and prey, yet capable of tenderness and affection towards their own kind, they are all, particularly the young, captivating and endearing creatures. Each species, with its own unique behavioural traits and environmental requirements, is at home on the continent, and Africa would be a much poorer place without them. All of Africa's carnivores need vast areas in which to roam and hunt, and defend their progenies, and by protecting these wild creatures we protect the very fabric of the environment in which so many other creatures exist.

# LION

*Lions are the only significantly social cats, and the bond enjoyed by prides offers not only communal defence, but the opportunity to breed successfully. Lionesses form the core of the pride, and by combining their strength they are often able to repel foreign males in the temporary absence of pride males, providing security to their cubs. Females produce cubs every three years, and pride lionesses often reproduce in synchrony. Prides hunt cooperatively, and feed communally on large kills. Males are ousted from the pride at around three years of age, and often join forces with other emigrant litter mates in search of their own territories.*

LEFT  Although tree raking is not an important male territorial signal, it is necessary for maintaining sharp claws, and cubs and adults alike frequently indulge in this practice.

OPPOSITE  Much of the play of juvenile lions seems to be focused on honing the muscle coordination and strength required to capture prey.

OVERLEAF LEFT  A lion's roar, predominantly emitted by males, is an efficient long-distance signal, and is interpreted differently by the various individuals or groups that hear it. For example, non-territorial or nomadic males tend to avoid or move away from an area where pride males are active.

OVERLEAF RIGHT  From scent traces, lions are able to determine information such as the sex, status and, in the case of a female, the reproductive condition of a lion that has previously passed through a territory. The lip curling and grimacing displayed by a lion while assessing another's scent is known as 'flehmen'.

LEFT, TOP TO BOTTOM An adult male lion consorts with a lioness. At first unresponsive, more often than not the female solicits mating, which happens fairly frequently over a period of three to four days; ovulation is only induced by repeated copulation.

ABOVE Copulation in lions, and indeed most of the large cats, is usually accompanied by some degree of aggression. In rare cases where a lioness refuses to mate with a dominant male, he may attack and even kill her.

OPPOSITE By consorting closely with a receptive lioness, a male lion secures his mating rights, although lionesses are not strictly monogomous when in oestrus, and may mate with other lions in the pride to ensure propagation.

OVERLEAF LEFT Cubs remain dependent on their mother's milk for up to eight or nine months, although they generally start feeding on meat from the age of about three months.

OVERLEAF RIGHT A cub's day consists of long periods of rest, interspersed with bouts play and feeding.

**OVERLEAF LEFT**  After a tiring morning of adventure and activity, there is nothing more comforting than snuggling up to mother, although this cub remains ever alert to a possible frolic.

**OVERLEAF RIGHT**  From about three months of age, cubs join lionesses on the prowl, and cover fairly large distances with them. However, the lionesses may leave them behind for considerable periods when they need to hunt.

**ABOVE**  Adult lions seldom climb trees, but cubs seem inexorably drawn towards them. Tree limbs, especially when bent and gnarled, provide a handy platform on which boisterous cubs play and scuffle with litter mates.

**RIGHT**  A pride's leisurely morning amble through the bush is often brought to a halt as cubs are beckoned by an easy-to-climb branch.

**OPPOSITE**  Even fairly large cubs are wont to climb trees, even though it means that they sometimes find themselves in precarious positions, usually resulting in an undignified meeting with the ground.

**ABOVE** After the initial feeding frenzy, where the soft, internal organs and main muscles are consumed, lions feed more slowly, eventually consuming almost every last piece of their prey, particularly of smaller species such as warthog.

**ABOVE** Larger cubs accompany adult lionesses on hunts, learning from an early age how to position themselves for a group hunt.

**OPPOSITE** Lion prides will sometimes follow descending vultures to investigate a carcass.

**OPPOSITE LEFT** Generally speaking, very large prey species are only infrequently killed by lions. However, lionesses hunting in big groups of five or more will often attempt to bring down a buffalo if the opportunity arises; during periods of drought, a weak adult buffalo in poor condition is easy prey.

**OPPOSITE RIGHT** A medium- to large-sized kill will occupy a small pride for a day or two, during which time their activities are almost exclusively devoted to eating and resting.

**ABOVE** Lion populations are generally well protected in Africa's many national parks and conservation areas, and, after the spotted hyena, lions are the most numerous predator on the continent.

**RIGHT** Some 80 per cent of a lion's day is spent resting or sleeping, and lions will yawn periodically before becoming active after a long rest. It is not unusual to see a mature or elderly lion without one or more of its canine teeth.

# LEOPARD

*Often referred to as the 'ultimate' cat, the graceful leopard epitomises feline perfection. Leopards are highly adaptable, occurring in a very wide range of habitats. Distributed throughout Africa and southern Asia, and found in rainforests, on mountains and in deserts, leopards have the largest range of all the world's big cats. In savanna regions, leopards occupy surprisingly small territories, but in harsh environments like the Kalahari have staggeringly large ranges. Although the leopard is smaller than the lion, this is no reflection on its courage and ferocity, as it is a powerful and lethal predator.*

LEFT Leopards have tremendous patience and, aided by the effective camouflage of their dappled coats, are stealthy and cunning hunters, capable of stalking prey to within very close range before charging or pouncing.

OPPOSITE Seldom active during the day, a leopard moves down to a local pan for a drink in the early evening. Leopards obtain most of their moisture requirements from prey, and are fairly independent of water, but will drink when it is available.

OVERLEAF LEFT Although leopards are found in arid environments, one typically associates them with areas of thick, lush cover.

OVERLEAF RIGHT As at home on the ground as they are in trees, leopards mainly utilise trees for resting, and stashing carcasses out of reach of other predators.

LEFT Both male and female leopards vigorously patrol and mark their territories. Long-lasting signals are left by scent-marking, by spraying urine on bushes and at the base of trees. Male leopards also vocalise loudly, with a rasping cough-like call. Males have distinctly thick, muscular necks and often develop dewlaps.

BELOW The territories of female leopards overlap more so than those of males, but interactions between neighbours are rare. Female cubs that disperse often establish territories adjacent to, and even incorporating some of their mother's territory.

OPPOSITE Subadult leopards begin to spend extended periods away from their mothers, and it is during this time that they make their first serious hunting attempts.

OVERLEAF LEFT Although superbly camouflaged – even during the day – leopards remain predominantly nocturnal.

OVERLEAF RIGHT Small ungulates constitute the bulk of a leopard's diet, but leopards are opportunistic hunters and prey on a diverse range of other, smaller creatures.

LEFT AND ABOVE Female leopards often abandon their cubs for several days at a time when they go off to hunt. Cubs are usually tucked away in a secure den or in the safety of a fallen tree. Young leopards can be surprisingly aggressive, and are often able to ward off inquisitive animals.

OPPOSITE For the first year of a cub's life it will share its mother's territory, but in the second year it becomes progressively more independent.

OVERLEAF LEFT Leopards living in arid areas are sometimes paler in colour and are usually slighter in build than those from more dense savanna areas.

OVERLEAF RIGHT The vibrassa whisker pattern is specific to individual leopards and can be used to 'fingerprint' them. Their long whiskers assist in determining whether or not a hole can accommodate them as they crawl into it in search of prey or to escape the heat of the day.

LEFT, ABOVE AND OPPOSITE An adult impala – favoured prey of leopards – is of similar weight to the cat, and it requires incredible strength to haul a kill of this size up into a tree. Leopards 'strangle' their larger ungulate prey, and, in an effort to avoid conflict with any scavenging predator, will stash a carcass in a tree, dragging it up to a safehold by the neck. Smaller prey is eaten on the ground immediately after capture, especially in desert environments where leopards seldom have the option of using trees. Leopards are not averse to eating rotting flesh, and will often revisit a hoarded carcass for some time after the kill.

OVERLEAF LEFT AND RIGHT A picture of peace and tranquility, the seeming lethargy of a leopard belies its true nature; at the close proximity of a foe – particularly the spotted hyenas that harass a leopard for its cache – the cat will become instantly ferocious.

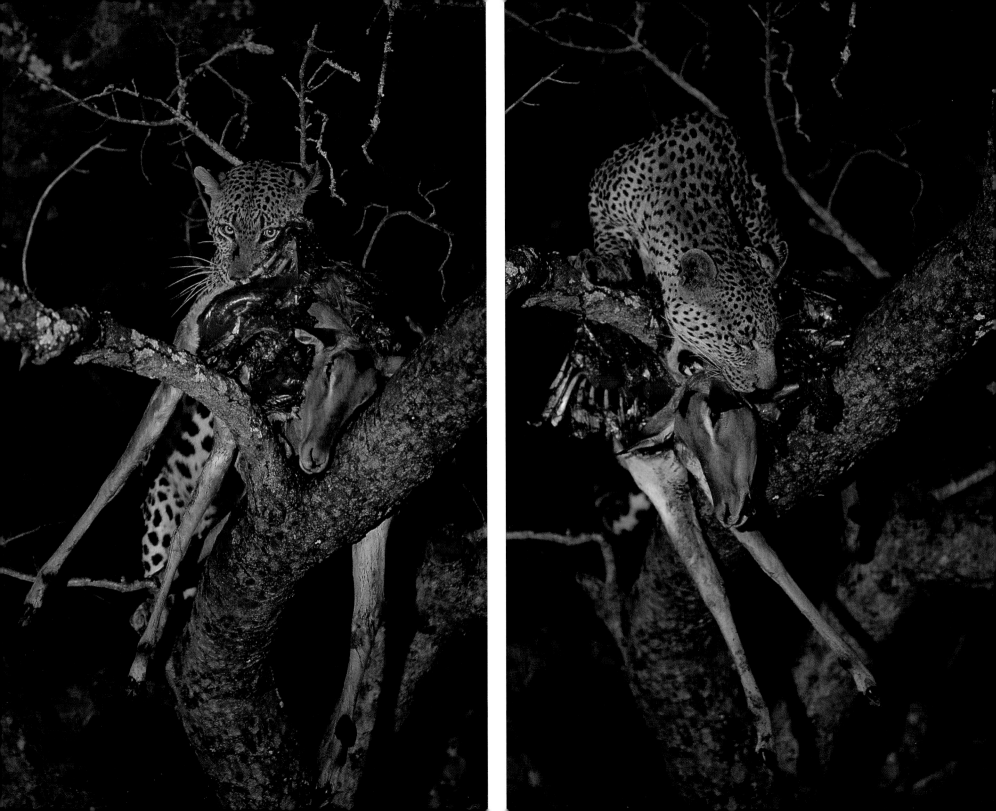

ABOVE LEFT AND RIGHT, AND OPPOSITE  The first leopards are thought to have been forest dwellers. Since then, the species has adapted to an exceptionally wide range of habitats. Leopards are nocturnal creatures, and will spend much of the day resting up in the dappled shade of long, dry grasses and trees where their spotted coats are well camouflaged, before setting out just before dusk to hunt, wandering for most of the night until a few hours before dawn.

OVERLEAF LEFT  A hungry leopard will prey on any animal when its primary food source (antelopes such as this impala) is scarce, and leopards have been observed eating creatures such as crocodiles, primates, birds, scorpions, snakes, lizards or insects.

OVERLEAF RIGHT  Leopards stalk and pounce on their prey, and are stealthy hunters, moving fluidly over obstacles without alerting unsuspecting prey to their presence.

# CHEETAH

*The cheetah is the sole modern day representative of the genus* Acinonyx, *derived from the Sanskrit and meaning 'spotted one'. Well known for their incredible short-distance bursts of speed, cheetahs stalk their prey to within close proximity, and then chase down the victim over about 300 metres (328 yards). This hunting technique in all probability developed in open plains-like environments – prime habitat for cheetahs, although they seem to be equally at home in fairly dense tree savanna regions. Unlike most other large cats, cheetahs hunt primarily during daylight hours. Throughout their distribution, there is a genetic uniformity in these sleek, lightweight and long-legged cats, and this has led zoologists to believe that all present-day cheetahs share a common ancestry.*

LEFT Two cheetah cubs realise the importance of remaining concealed in an environment where they may easily fall prey to other predators.

OPPOSITE Cheetahs frequently scan the environment for possible prey from elevated areas such as in trees or on termite mounds, where they often mark the territory with urine or faeces.

OVERLEAF LEFT Because they hunt predominantly by sight, and in order to avoid other large predators such as lions and spotted hyenas, cheetahs are mainly active during the day, particularly during the cool of the early morning and evening. Siblings typically stay close together, particularly in arid, open areas such as the Kalahari.

OVERLEAF RIGHT A cheetah's fragile frame does not facilitate protection from other large, scavenging carnivores, and in open areas they will quickly consume the protein-rich large muscle tissue of the hindquarters and the nutrient-rich internal organs first; if cover is available, a cheetah will make use of it by dragging the carcass out of sight.

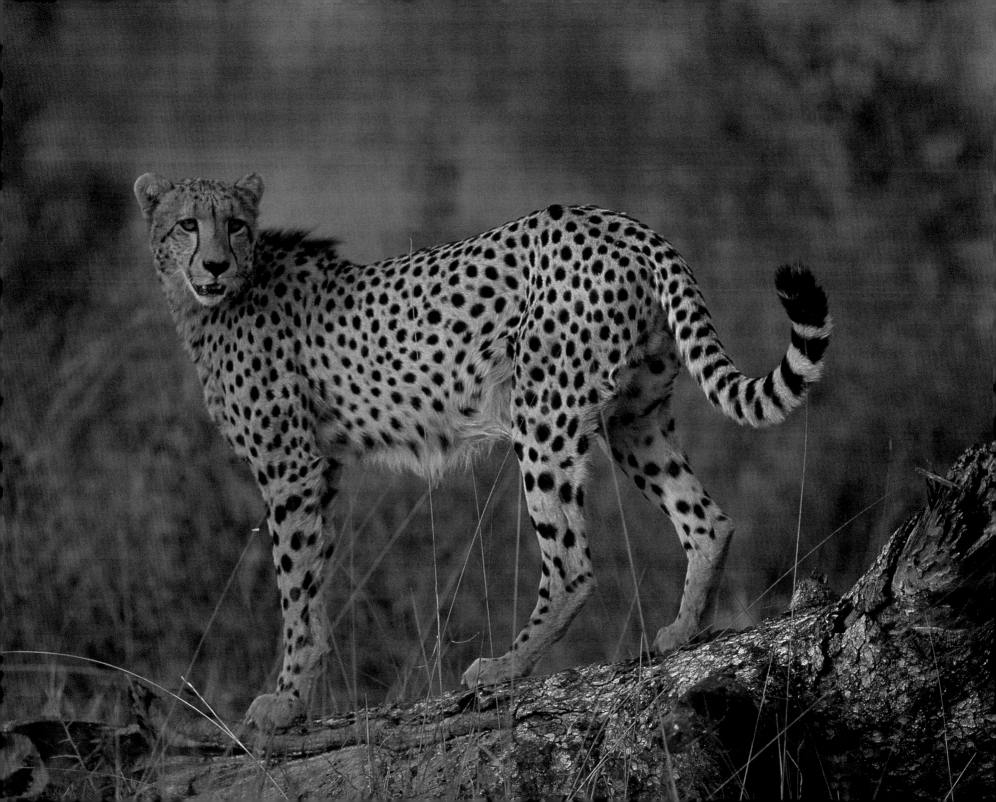

OPPOSITE Single females seldom attempt to bring down prey much larger than a small antelope. Males, on the other hand, are bigger and more heavily built, and usually have one or two coalition partners, ensuring successful hunts of larger ungulates such as gemsbok and even wildebeest.

OVERLEAF LEFT Cheetahs seldom seem truly at ease, even when resting, as they regularly lift their heads or sit up to scan the area for potential danger or prey.

OVERLEAF RIGHT The king cheetah is not a separate species, but merely represents an aberrant colour form of the cheetah brought about by a recessive gene; unlike the common cheetah, though, its magnificent pelt consists of dark spots that combine and run into attractive bars and streaks.

ABOVE Cheetahs are widespread in Africa and southern Asia, mainly inhabiting open savanna and semi-desert environments. Cheetahs are the only cats that do not possess retractible claws, making their spoor relatively easy to distinguish.

RIGHT As with most predators, cheetahs are opportunistic hunters, and will usually prey on any animal that wanders within easy stalking or sprinting range.

THIS PAGE A hunting cheetah employs different strategies when approaching prey, depending on terrain, species, and behaviour of the animal. Usually, though, a cheetah will seek out prey from a suitable vantage point, selecting the least vigilant animal on the fringe of a group. The cheetah will stalk under cover if it is available; if not, it is just as comfortable moving across open ground, creeping forward when the animal turns away or lowers its head to feed, and freezing when it looks up. A cheetah will give up a stalk if it is seen by prey within about 100 metres (110 yards), as it depends on the element of surprise for success. Although a cheetah's top speed is an incredible 112 kph (70 mph), it can only sprint for about 300 metres, or 328 yards, before rising body temperature and oxygen deficit force it to quit.

OPPOSITE Cheetahs will display aggressively, not only towards their own kind, but also towards other animals. However, the display is pure bluster – when they encounter any of the larger predators, cheetahs will invariably turn tail and run.

PREVIOUS PAGE LEFT When cheetahs feed, they lift their heads at regular intervals to scan the area nervously, making sure that the chase has not attracted other predators.

PREVIOUS PAGE RIGHT Female cheetahs raise their cubs – blind and helpless at birth – on their own, frequently changing dens as the cubs are very vulnerable.

LEFT AND ABOVE A female cheetah that is about to become receptive is located by males from the smell of her urine, and mating usually takes place after an aggressive courtship.

OPPOSITE While the bond between cheetah siblings is not as intense as that between lions, and they seldom lie in contact with one another, litter mates will often stay together for some time after separating from their mother.

OVERLEAF LEFT Cheetah cubs are fair game for a number of predators, most notably, lions, and in the Serengeti, less than half the number of cheetahs born survive their first three months.

OVERLEAF RIGHT The cubs are half grown by their sixth month and are fully weaned in the following two months. Once weaned they take an increasing interest in learning how to hunt.

ABOVE A blood-stained face is a certain indication that a cheetah has recently killed. While small prey is completely consumed, with larger prey the cheetah usually abandons the carcass once it has had its fill, leaving most of the skeleton and skin for scavengers, and making a cheetah kill fairly easy to identify.

ABOVE Trying to conceal several inexperienced cubs in a hunt is very difficult and leads to many missed opportunities for the hunting adult.

OPPOSITE By the time cubs are a year old, they are able to kill for themselves. The mother will encourage cubs by catching small prey on which they can practise their hunting skills.

# SERVAL

*Fawn-coloured servals, tallest of the smaller cats, are specialist hunters of grassland-dwelling animals such as rodents and birds. Their predatory successes may be reliant on their acute hearing, which helps them to pick up the rustle of mice and rats as they scurry through the grasses, and their long legs, which enable the cats to leap high into the air (also improving hearing range) before pouncing on unsuspecting prey in a flurry of black bands and spots. Often found in forest fringes, and usually near water, servals have been known to catch birds mid-air, and to stalk water birds and amphibians in the water.*

LEFT Easily identifiable by the distinct markings on the backs of its ears, the serval is usually a solitary, nocturnal animal.

OPPOSITE A hunting serval will stand stock-still while listening for a faint rustling in the grass that indicates the presence of prey. A serval sometimes leaps as far as two metres (six and a half feet), to land precisely on its quarry. It then holds its prey down with one forepaw while batting viciously at it with the other, until it is able to part the grass stems and seize the victim in its mouth. This stalk-and-pounce procedure, however, is not as efficient with larger prey.

OVERLEAF LEFT On rare occasions, a serval will tackle dangerous prey such as a puffadder, where its nimble leaps and ability to change direction at great speed help the cat to avoid being bitten.

OVERLEAF RIGHT Although predominantly solitary, male and female servals sometimes forage together, presumably as a prelude to mating. Kittens are hidden in clumps of vegetation, and are carried away when danger threatens.

# CARACAL

*Caracals are mainly nocturnal, territorial and solitary, and males and females associate only for mating purposes. Females separate from their kittens as soon as the young are able to fend for themselves. Usually terrestrial, they are nevertheless adept tree climbers, making full use of their powerful dew-claws in this pursuit. Equipped with remarkably powerful hindquarters, caracals are swift and strong. They are primarily hunters and are generally averse to taking carrion, although they have been recorded doing so. Distributed throughout Africa's arid zones and dry savannas, caracals remain heavily persecuted in stock-farming areas, even though their demise often leads to increasingly high populations of more destructive species.*

LEFT, TOP TO BOTTOM  A caracal's face, with its black facial markings, is strikingly attractive. The cat's distinctively black-backed and tufted ears dramatically alter the caracal's expression, presenting aggression when they are swung backwards.

OPPOSITE  Caracals prey on a wide range of small- to medium-sized mammals and birds. Dry grass and bushes provide the reddish-fawn-coloured animal with ample camouflage, from which it stalks its prey to within metres. Caracals, like their other, bigger cat cousins, also suffocate larger prey with a throat bite, and they are as adept at taking down an antelope twice their weight, as they are at rock jumping to pursue hyraxes.

OVERLEAF LEFT  Caracals of both sexes scent-mark in their territories by urine-spraying, and leave visual and scent signals when they sharpen their claws on wood or bark. Faeces is either buried or left exposed on tracks and pathways.

OVERLEAF RIGHT  It is thought that caracal females bear litters under substantial cover, particularly in disused aardvark holes, which they line with hair or feathers from kills.

# SPOTTED HYENA

*Spotted hyenas are highly social, fascinating and indeed enchanting creatures, capable of astounding hunting feats, and exhibiting complex and intriguing social behaviour. The massive head and neck, powerful jaws and hugely muscled shoulders and hindquarters clearly show that this creature is built for power, and it uses all these attributes for latching onto and pulling down large ungulate prey, as well as for tearing apart bigger carcasses. Having teeth and jaws reminiscent of a crushing mill, spotted hyenas are capable of splintering even very large bones, thus accessing the highly nutritious marrow contained within. The image of these animals being lowly, cowardly scavengers has almost faded into myth as they are now recognised as true and highly successful predators.*

LEFT  Spotted hyenas are born in dens ranging from converted aardvark holes to natural caves and road culverts, or, more often, disused porcupine warrens. The den, which is moved periodically, forms the focal point of clan activity, and even if the hyenas are not at the den itself, most adults rest in its vicinity during the day.

OPPOSITE  The cubs live on milk alone for six to nine months, only visiting kills when they are about a year old, although they can digest meat from a very young age. Cubs remain in the vicinity of the den almost entirely until they are weaned at about 14 to 18 months – the longest period of time that any carnivore is dependent on maternal milk.

OVERLEAF LEFT AND RIGHT  Spotted hyena cubs are often left behind at the den for many days while their mothers are out foraging. Cubs, like any other young animals, will progressively begin to explore the surrounding area, playfully biting branches of small bushes and carrying pieces of wood and old bones around.

THIS PAGE Spotted hyenas are capable of tremendous feats of endurance, both in covering enormous distances while foraging, and in their ability to stay with a chase, following selected prey for up to several kilometres until the victim is so exhausted it can no longer keep running. This hunting technique is extremely successful, and in the Kalahari it has been documented that there is more probability of a spotted hyena bringing down a gemsbok that has run from it, than one that has stood its ground and defended itself.

OPPOSITE While hyenas are successful predators, in times of little they will seek out carcasses, settling for even the most meagre pickings in an effort to satisfy nutrient requirements.

PREVIOUS PAGE LEFT Hyena cubs are very playful, running and chasing each other and pestering the adults by climbing over them and biting. Scent-marking and ritualised social behaviour begins when the cubs are only about four to six weeks old, even though their scent glands are still inactive.

PREVIOUS PAGE RIGHT Spotted hyena cubs have an extraordinary capacity to consume milk, and suckle for extended periods, especially after the mother has been away for several days.

LEFT Unlike brown hyenas, spotted hyenas rarely carry food back to the den, although adults are often seen carrying carrion away from kills; this is usually consumed at some quiet location before the hyena returns to the den.

ABOVE Females usually bear two young, and an excess of two cubs in a den will more often than not be the progeny of more than one female.

OPPOSITE Spotted hyenas differ from other social carnivores in that the larger-bodied females are dominant. Females compete for rank, and although they raise cubs in communal dens, they do not cross-suckle offspring. Males play no parental role, and only a few privileged individuals are permitted near the den.

TOP LEFT, LEFT AND ABOVE Spotted hyenas have a relatively long gestation period of four months, resulting in the birth of fairly precocious offspring; cubs are born with open eyes, fully developed incisors and canines, and capable movement of the forelimbs. Although covered in black hair, the cub's coat soon turns fair and becomes spotted. Large cubs are often seen begging milk from adult females that are suckling cubs and are thus probably not their mothers. These attempts are invariably rebuffed.

LEFT Adult females will carry cubs to a communal den some 10 days after giving birth in a separate burrow.

OPPOSITE Cubs are seemingly adventurous and carefree, but will bolt into the safety of the den at the slightest sign of danger.

# BROWN HYENA

*These creatures are usually more common in a particular area than the infrequent sightings of them might suggest. Almost completely nocturnal, they are sometimes observed on cool afternoons, as they begin their foraging rounds after lying up under a bush for most of the day. Brown hyenas live in clans of about eight to 10 individuals and share a communal den; however, unlike spotted hyenas, brown hyenas predominantly forage alone. They are almost exclusively scavengers and, under the mantle of their shaggy coats, walk vast distances at night in search of the merest morsel of food. Brown hyenas do not compete aggressively with other large carnivores for the spoils of a kill, but will lurk, waiting for predators to move away from a kill before one or two of them move in for the scraps, aggressively chasing off jackals that are invariably the first at the scene.*

LEFT Probably in order to reduce the likelihood of aggressive interactions with another large carnivores, brown hyenas usually carry the largest possible piece of carcass remains away from the kill location. They will often stash these portions and return for more.

OPPOSITE Brown hyenas do not necessarily make the long trip back to their group's den at the end of a night's foraging. If they find a rich food source, they make a temporary camp by scooping out a hollow in the soil under the cover of long grass, bushes or trees.

# WILD DOG

*Of all Africa's predators, none has been more abused or misunderstood than the wild dog. Persecuted throughout Africa, even conservation organisations were once convinced that wild dogs would eliminate smaller antelope species and actually requested that wild dogs be shot on sight as they were 'too ruthless and too horrible to preserve'. It is little wonder, therefore, that these animals are today an endangered species. One single factor lies behind the wild dog's survival: its efficiency as a predator. Wild dogs depend on their own kills for sustenance, and exhibit a very high rate of hunting success. Singling out the young and vulnerable from the herd, the pack will give chase before mobbing the animal, bowling it over and consuming it in a matter of minutes.*

LEFT Wild dogs live in packs that can number from two to 50 individuals, fluctuating greatly with the birth of very large litters, frequent mortalities and dispersing same-sex groups. The pack is headed by a dominant male and female, who, in the vast majority of cases, account for all the breeding activity.

OPPOSITE Wild dogs are unusual in that adult males remain in the pack to become helpers, while the females disperse from the pack.

OVERLEAF LEFT Although subordinate members of the pack avoid aggression by taking up a submissive posture in the company of more dominant dogs, severe facial injuries – presumably a result of fighting – are occasionally noted.

OVERLEAF RIGHT Wild dogs are distributed throughout Africa's savanna and arid regions, and while seldom found in forested areas, they do occur in dense scrub.

ABOVE Wild dogs have handsomely blotched coats of black, white and brown, and each individual can be identified by its unique patterning. They communicate vocally and enjoy close contact, especially while resting in 'heaps'.

OPPOSITE Pups are born in holes in the ground, very often in disused aardvark or warthog burrows in termite mounds. Unlike most other large carnivores, wild dogs are seasonal breeders, and the alpha female usually produces a litter of pups in the mid-winter months.

THIS PAGE AND OPPOSITE For about three to four weeks after whelping, before pups emerge and begin taking solid food, the mother stays at the den while the pack goes hunting and often will not let the other dogs near the burrow entrance. Pups are weaned as early as five weeks, when they will eat food disgorged by returning pack members. Dogs returning from a hunt also regurgitate food for the mother and babysitters that remain at the den.

OVERLEAF LEFT When prey is sighted, the dogs will sometimes sneak up on it from cover of bushes. If the quarry scatters, as the most common prey, impala, usually does, the pack splits up and multiple kills may be made. Although reputed to be efficient pack hunters, it seems as if most kills in dense bush are made by a single dog that leads the pack, relentlessly chasing down the prey, and nipping at its underparts until the animal weakens and falls.

OVERLEAF RIGHT Once a kill has been made, the juveniles and subadults rather than the adults are allowed to feed first. Carcasses are consumed quickly, without the snarling and bickering that accompanies lion and spotted hyena kills.

LEFT Pups lose their rotund shape early, and begin to assume adult features from seven weeks.

ABOVE AND OPPOSITE All members of the pack are remarkably attentive toward the pups, and regularly pick them up – often against their will – and deposit them at the entrance to the den if they wander away from it. Adults and subadults alike will help to maintain the den area, digging and enlarging burrows and keeping the surrounds clean. When pups are about two months old the pack abandons the den.

OVERLEAF Wild dogs show little fear of other carnivores, and although they alarm-bark and generally move out of the way of lions, they have been known to 'tree' a leopard. Relations with spotted hyenas are highly competitive because hyenas regularly try to steal their kills. Despite the obvious size difference, hyenas are often seen off by the dogs, which will mob them relentlessly if they become bothersome.